Forms of Contract

Adjudication rules

The Grey Book

Third edition
2004

**Published by
Institution of Chemical Engineers (IChemE)
Davis Building
165–189 Railway Terrace
Rugby
Warwickshire
CV21 3HQ, UK**
IChemE is a Registered Charity in England and Wales

Copyright © 2004
Institution of Chemical Engineers

ISBN 0 85295 470 0

First edition 1998
Second edition 2001
Third edition 2004
Reprinted 2005

Printed in the United Kingdom by Hobbs the Printers Ltd, Totton, Hampshire

Preface

These Rules have been produced by the Institution of Chemical Engineers (IChemE) in response to the statutory right to adjudication introduced to construction contracts for work in England and Wales and in Scotland by the Housing Grants, Construction and Regeneration Act 1996, and for work in Northern Ireland by the Construction Contracts (Northern Ireland) Order 1997. They are intended to be used in conjunction with IChemE's Forms of Contract for work covered by the provisions of the Act or the Order. Clauses on statutory adjudication are incorporated in IChemE's Forms of Contract.

The aim of adjudication is to reach a fair, rapid and inexpensive resolution of referable disputes arising under the Contract.

This third edition has been prepared in the light of decisions of the Courts and in response to comments received from users.

August 2004

Acknowledgements

These Rules have been produced by the Disputes Resolution Panel of the Institution of Chemical Engineers (IChemE). Members of the Disputes Resolution Panel are:

Henry Rowson (Chairman)

Gordon Bateman

John Challenger

Graham Rutherford

David Wilson

David Wright

IChemE thanks Paul Buckingham, Clifford Chance LLP, for his input and advice on preparing this new edition.

IChemE acknowledges its debt to the Institution of Civil Engineers, from whose Adjudication Rules some elements of these Rules have been derived.

Adjudication and the Housing Grants, Construction and Regeneration Act 1996 and the Construction Contracts (Northern Ireland) Order 1997

Adjudication is embodied in Part II of the Housing Grants, Construction and Regeneration Act 1996. It is a quick method of resolving disputes, providing a decision which is binding on the parties, but which is not necessarily final. If they choose to litigate or arbitrate or otherwise agree a resolution of their dispute at a later date then they are free to do so. In the short term the decision of the adjudicator must be adhered to, and is enforceable in the Courts. While primarily aimed at problems within the building and civil engineering industries, the Act has potentially wider impact. The Act implies terms into contracts for 'construction operations'.

The Act applies only to England and Wales and to Scotland. The Construction Contracts (Northern Ireland) Order 1997 (the Order) applies very similar rights and obligations to Northern Ireland.

The definition of 'construction operations' in the Act and the Order is, however, sufficiently wide to include many contracts that have little or nothing to do with the civil engineering or building industries. The definition will certainly extend to many contracts and subcontracts in the chemical and process industries. This means that in any contract within the process industries that includes construction operations as defined by the Act, the right to adjudication in the event of a dispute is overlaid onto the dispute resolution methods provided in the contract.

Very briefly, 'construction operations' are defined in the Act and the Order as work within the UK on buildings and structures, works at or below ground level—such as walls, pipelines, railways and reservoirs, the installation of equipment in buildings or structures, and site preparation work such as foundations and earthmoving. They also include repair and maintenance as well as construction, alteration and demolition. Even painting is included, as are design and other consultancy work associated with construction operations. However, 'construction operations' do not include (among others) the mere supply of equipment or some work involving plant/machinery/supporting steelwork on sites where the main activity is water/effluent treatment or the production, storage or processing of chemicals, pharmaceuticals, oil, gas, steel, or food products. Where a contract is both for 'construction operations' and those matters which are excluded from that defined term, the implied terms apply only to disputes associated with the 'construction operations'. Hence there is a possibility of different parts of the works being or not being potentially subject to adjudication.

This is not the right place for an exhaustive discussion of what adjudication means or implies. The important point is that if the Act's or the Order's provisions apply they cannot be avoided if either party to the Contract chooses to invoke them. The probability is that the right to adjudication of at least some disputes will apply in most UK main Red, Green or Burgundy Book contracts, and also in those subcontracts of any tier which include any 'construction operations'. For this reason the Brown Book (subcontract for civil engineering works) provides for adjudication to be available for any dispute under the contract.

Those involved with a dispute which is potentially referable to adjudication should consider the practicalities of the short time within which the procedures must be completed. The facts of most disputes arising from process plant contracts will be complex. The Act and the Order only allow an Adjudicator some four weeks at most (after allowing time for the preliminaries, etc) to gather his information and then assess that information and produce his decision. This could be almost impossibly short. The information could be considerable and conflicting, and would take a considerable time for the Adjudicator to assimilate and understand.

It should be noted that the IChemE Forms of Contract have for many years recognised that disputes require different methods of resolution depending on their complexity. It is assumed that most parties to disputes will wish to avoid litigation in the courts due to the extensive time-scales and costs which will almost certainly be incurred. The Forms of Contract have all retained a facility to appoint an Arbitrator and, although thought to be less costly than resorting to the courts, this can be a lengthy and complex procedure which should only be used for disputes with relatively high complexity, albeit that the IChemE Arbitration Rules contain provisions for short form arbitrations in appropriate circumstances.

The Red, Green, Burgundy, Brown and Yellow Books also provide the facility for the appointment of an Expert to resolve matters of a mainly technical, contractual and quantum nature where the solution to a dispute is complex and finality is sought in the award. Expert determination is applicable to a range of possible disputes including those which, because of the exemptions from the definition of 'construction operations', would not be covered by the Housing Grants, etc Act.

Adjudication is a procedure in which problems of a relatively minor nature can be resolved quickly, thus lessening the likelihood of a general collapse of working relationships or delays to the project programme, and a consequential escalation of the dispute to other areas of a contract. At subcontract level and minor works level adjudication is more feasible than for main contracts.

There is a current tendency in at least some areas of building construction to extend the use of adjudication to relatively large and complex disputes. To date this approach has not found favour in the process industries where there is a clear preference for 'getting it right' rather than 'getting there fast'.

Contents

Page

Rules for the conduct of adjudications

1. General principles and liabilities

1.1 The adjudication shall be conducted in accordance with the edition of IChemE's Adjudication Rules current at the date of issue of a Notice to refer a dispute to adjudication as defined in Rule 2.1. If a conflict arises between these Rules and the Contract then these Rules shall prevail.

1.2 The Adjudicator shall be a named individual and shall act impartially.

1.3 In making a decision, the Adjudicator may take the initiative in ascertaining the facts and the law. The adjudication shall be neither an expert determination nor an arbitration but the Adjudicator may rely on his own expert knowledge and experience.

1.4 The Adjudicator's decision shall be binding until the dispute is finally determined by legal proceedings, by arbitration (if the Contract provides for arbitration or the Parties otherwise agree to arbitration) or by agreement. The Parties may agree to accept the decision of the Adjudicator as finally determining the dispute.

1.5 The Parties shall implement the Adjudicator's decision without delay whether or not the dispute is to be referred to legal proceedings or arbitration.

1.6 These Rules shall be interpreted in accordance with the law of the country where the Site is situated.

1.7 The Adjudicator shall not be appointed arbitrator in any subsequent arbitration between the Parties under the Contract, nor shall he be appointed as mediator between the Parties. Neither Party may call the Adjudicator as a witness in any legal proceedings or arbitration concerning the subject matter of the adjudication.

1.8 The Adjudicator shall not be liable for anything done or omitted in the discharge or purported discharge of his functions as Adjudicator and any employer, employee or agent unless the act or omission is in bad faith, and any employee or agent of the Adjudicator shall similarly be protected from liability. The Parties shall save harmless and indemnify the Adjudicator and any employer, employee or agent of the Adjudicator against all claims by third parties in respect of the adjudication and in respect of this the Parties shall be jointly and severally liable.

1.9 The Parties shall agree to the exclusion of IChemE, its servants and agents from liability to any Party for any act or omission or misconduct in connection with any selection or appointment made or any adjudication performed in consequence of an application for the appointment of an Adjudicator.

1.10 Any reference in IChemE's Forms of Contract to Adjudication Procedures shall be deemed to be a reference to these Rules.

2. Definitions and interpretation

2.1 Unless the context otherwise requires, the following expressions shall have the meanings hereby assigned to them.

'Adjudicator' means the person named as such in the Contract or appointed in accordance with these Rules.

'Contract' means the contract or the agreement between the Parties which contains the provision for adjudication.

'IChemE' means the Institution of Chemical Engineers.

'Notice' means a written notice of adjudication in the format given in Annex A.

'Party' means a party to the Contract.

The 'Act' means the Housing Grants, Construction and Regeneration Act 1996.

The 'Order' means the Construction Contracts (Northern Ireland) Order 1997.

2.2	Any term defined in the Contract shall have the same meaning in the context and interpretation of these Rules.

2.3   Annexes A to D shall be considered to be part of these Rules.

2.4   'Day' shall mean a calendar day.

In England and Wales and in Scotland where an act is required to be done within a specified period after or from a specified date, the period begins immediately after that date. Where the period would include Christmas Day, Good Friday or a day which under the Banking and Financial Dealings Act 1971 is a bank holiday in England and Wales or in Scotland, as the case may be, that day shall be excluded.

In Northern Ireland where an act is required to be done within a specified period after or from a specified date, the period begins immediately after that date. In all other respects the period shall be reckoned in accordance with the Order.

2.5   The preface and introductory notes do not form part of these Rules.

2.6   In these Rules the singular shall include the plural and the plural the singular except where the context requires and the words 'he', 'him' and 'his' shall be taken to mean 'she', 'her' and 'hers' where appropriate.

3.   Issuing of the Notice

3.1   The Party wishing to refer to adjudication any dispute arising under or in connection with the Contract may give a Notice at any time to that effect to the other Party. The Notice shall include:
(a)   the details and date of the Contract;
(b)   the issues which the Adjudicator is to be asked to decide; and
(c)   the nature and extent of the redress sought.

4.   Appointment of the Adjudicator

4.1   When an Adjudicator has either been named in the Contract or agreed prior to the issue of the Notice, the Party issuing the Notice shall at the same time send to the Adjudicator a copy of the Notice with a request for confirmation, within four days of the date of issue of the Notice, that the Adjudicator is able and willing to act.

4.2   When an Adjudicator has not been so named or agreed, the Party issuing the Notice may include with the Notice the name(s) of one or more persons with their addresses who are willing to act, and are acceptable to the referring Party, for selection by the other Party. Provided that one of these is acceptable to him, the other Party may select and notify the referring Party and the selected Adjudicator within four days of the date of issue of the Notice.

4.3   Failing agreement to nominate an Adjudicator under Rule 4.1 or 4.2 above, either Party may within a further three days request the President for the time being (or a Past President) of IChemE, to appoint an Adjudicator. Such request shall be in writing in the form of application given in Annex C, accompanied by a copy of the Notice and the appropriate fee as referred to in Annex D.

A copy of the request, together with the supporting documentation, shall be sent at the same time to the other Party.

4.4   The Adjudicator shall be appointed on the terms and conditions set out in the form of agreement as given in Annex B and shall be entitled to be paid a reasonable fee together with his expenses. The Parties shall sign the agreement within seven days of being requested to do so.

4.5   If for any reason the Adjudicator is unable to act, either Party may require the appointment of a replacement Adjudicator in accordance with the procedures in this Rule 4.

5. The referral

5.1 Within two days of receipt of confirmation from the Adjudicator under Rule 4.1, or notification of selection of the Adjudicator under Rule 4.2, or appointment of the Adjudicator under Rule 4.3 the referring Party shall send to the Adjudicator, with a copy to the other Party, a full statement of his case which should include:
(a) a copy of the Notice;
(b) a copy of any adjudication provision in the Contract; and
(c) the information upon which he relies, including supporting documentary evidence.

5.2 The date of referral of the dispute to the Adjudicator shall be the date upon which the Adjudicator has received all the documents and information referred to in Rule 5.1 from the referring Party, and he shall notify both Parties accordingly.

6. Early termination of the Adjudicator's appointment

6.1 Following his appointment the Adjudicator may resign at any time on giving notice in writing to the Parties.

6.2 The Adjudicator must resign where the dispute is the same or substantially the same as one which has previously been referred to adjudication, and a decision has been taken in that adjudication.

6.3 If the Adjudicator resigns for any reason other than that stated in Rule 6.2
(a) the referring Party may serve a fresh notice under Rule 3 and thereafter the provisions of Rule 4 shall apply to the appointment of a new adjudicator; and
(b) if requested by the new Adjudicator and insofar as it is reasonably practicable, the Parties shall supply him with copies of all documents which they had made available to the previous Adjudicator.

6.4 The Parties may at any time agree to revoke the appointment of the Adjudicator.

6.5 Where
(a) the Adjudicator resigns in circumstances as described in Rule 6.2, or
(b) a dispute varies significantly from the dispute referred to him in the Notice and for that reason he is not competent to decide it and resigns in accordance with Rule 6.1, or
(c) the Adjudicator's appointment is revoked in accordance with Rule 6.4

the Adjudicator shall be entitled to the payment of such reasonable amount as he may determine by way of fees and expenses reasonably incurred by him, including those of any legal or technical advisor engaged under Rule 7.5. The provisions of Rule 8.6 shall apply to such fees and expenses.

7. Procedure for the adjudication

7.1 The method used by the Adjudicator to conduct the adjudication shall be at his sole discretion. He shall establish the timetable subject to any limitation that there may be in the Contract or the Act or, as may be applicable, the Order and advise the Parties accordingly.

7.2 The Adjudicator shall not be required to observe any rule of evidence, procedure or otherwise of any court.

7.3 The Adjudicator shall consider the matters set out in the Notice, together with any other matters which the Parties and the Adjudicator agree should be within the scope of the adjudication.

7.4 The Adjudicator shall have the power to:
(a) decide any question of interpretation of the Contract between the Parties that is relevant to the dispute;
(b) ask for further written information;
(c) meet and question the Parties either separately or together;
(d) require any Party to make available for inspection any premises or item pertinent to the dispute, such inspection to be in the presence of representatives of the Parties should they elect to take part;
(e) require the production of documents or the attendance of people whom he considers could assist;
(f) direct that any property of or under the control of either of the Parties be preserved and, if necessary, stored;

(g)   call meetings to be attended by representatives of both Parties at locations and times chosen by the Adjudicator after consultation with the Parties, provided that if any Party unreasonably objects to the Adjudicator's proposed location, he may proceed to hold the meeting and advise the objecting Party of the outcome;

(h)   impose reasonable time limits on the Parties for replies to requests by him for information, and in the event of a Party failing to respond in accordance with the time limits imposed, to proceed with the determination in accordance with evidence available, and to draw such inferences as seem to be appropriate arising from such failure;

(i)   set times for (b) to (g) and similar activities;

(j)   proceed with the adjudication and reach a decision even if a Party fails:
  (i)   to provide information;
  (ii)   to attend a meeting;
  (iii)   to take any further action requested by the Adjudicator;

(k)   give a decision on different aspects of the dispute at different times;

(l)   issue such further directions or take such actions as he considers to be appropriate.

7.5   In addition, the Adjudicator may engage one or more advisors on any matter to assist in reaching a decision, provided that before making such appointment the Parties shall be given notice of such intention and where possible an indication of likely costs. In spite of the appointment of advisors, all decisions in the case are the Adjudicator's alone.

7.6   The Adjudicator may open up, review and revise any decision (other than that of an Adjudicator unless agreed by both Parties) made under or in connection with the Contract and which is relevant to the dispute. He may order the payment of a sum of money, or other redress. In particular, he may open up, review and revise any decision taken or certificate given by any person referred to in the Contract unless the Contract states that the decision or certificate is final and conclusive.

7.7   If at any time after the appointment of an Adjudicator the Parties reach an agreement on all or any part of the matter(s) under dispute, they shall send to the Adjudicator a statement to that effect, requesting that the Adjudicator should terminate the relevant part of the adjudication and render accounts of the fees and expenses due for payment by the Parties.

7.8   The Parties and the Adjudicator shall at all times maintain the confidentiality of the adjudication and shall endeavour to ensure that anyone acting on their behalf or through them will do likewise, except as may otherwise be agreed.

8.   The decision

8.1   The Adjudicator shall reach his decision within twenty-eight days of referral, or such longer period as is agreed by the Parties after the dispute has been referred. The period of twenty-eight days may be extended by up to fourteen days with the consent of the Party by whom the dispute was referred. The Adjudicator may reach a decision on different aspects of the dispute at different times.

8.2   The Adjudicator may award interest on any sum awarded to a successful Party to compensate that Party for any delay in the receipt of payment of such sum or any part of it, such interest to be applied at the rate (if any) prescribed in the Contract for wrongful failure to pay, unless otherwise agreed by the Parties. If there is no rate so prescribed then the Adjudicator may decide the rate that shall apply.

8.3   Should the Adjudicator fail to reach his decision and notify the Parties in due time, either Party may give both the Adjudicator and the other Party seven days notice of his intention to refer the dispute to a replacement Adjudicator to be appointed in accordance with the procedures in Rule 4.

8.4   If the Adjudicator fails to notify the Parties that he has reached his decision in due time, but subsequently does so notify the Parties before the date of appointment of a replacement Adjudicator under Rule 8.3, then that decision shall nevertheless be effective.

If the Parties are not so notified then no decision made by the Adjudicator shall be of any effect and he shall not be entitled to any fees or expenses. However, the Parties shall be responsible for the fees and expenses of any legal or technical advisor appointed under Rule 7.5 subject to the Parties having received such advice.

8.5   The Parties shall bear their own costs and expenses incurred in the adjudication.

8.6 The Parties shall be jointly and severally liable for the Adjudicator's fees and expenses, including those of any legal or technical advisor engaged under Rule 7.5, and unless the Adjudicator directs otherwise shall pay all such fees and expenses promptly and in equal shares.

8.7 At any time until seven days before the Adjudicator is due to reach his decision, he may give notice to the Parties that he will deliver it only on full or part payment of his fees and expenses. Either Party may then pay these costs in order to obtain the decision and recover the other Party's share in accordance with Rule 8.6 as a debt due.

8.8 The Adjudicator shall give reasons for his decision upon delivering it, unless otherwise agreed by the Parties.

8.9 The Adjudicator may, within 14 days of a mistake or error coming to his attention, correct any such mistake or error arising from an accidental slip or omission in a decision.

8.10 The Parties shall be entitled to seek to enforce the decision whether or not the dispute is to be referred to legal proceedings or arbitration. No party shall be entitled to raise any right of set-off, counterclaim or abatement in connection with any enforcement proceeding.

8.11 In the event that the dispute is referred to legal proceedings or arbitration, the Adjudicator's decision shall not inhibit the court or Arbitrator from determining the Parties' rights or obligations anew.

8.12 The Adjudicator shall inform the Parties if he intends to destroy the documents which have been sent to him in relation to the adjudication, and shall retain such documents for a further period at the reasonable request of either Party.

Annex A—Form of Notice of adjudication

Referring Party's reference number   *Referring Party's address*

                                                        Date

Name and address of responding Party

Sirs,

Notice of adjudication: re *Contract reference and date*

We hereby give notice that we require the following dispute to be referred to adjudication.

Issues we consider to be in dispute are as follows:

. . . .

. . . .

Th redress sought is as follows:

. . . .

. . . .

The following person(s) has agreed to act and is proposed for your consideration:

Name and address of potential Adjudicator(s)

You are required to agree in writing to the appointment of (one of) the person(s) named above within four days of the date of this Notice.

Failing receipt of such written agreement, we shall request the President (or a Past President) of the Institution of Chemical Engineers to appoint an Adjudicator.

Yours faithfully,

For and on behalf of the referring Party

Annex B—Form of Adjudicator's agreement

Adjudicator's reference number   *Adjudicator's address*

                                                                    Date

Name and address of the Parties

A

B

Sirs,

I have been appointed by *(agreement between the Parties/the Institution of Chemical Engineers (IChemE))*[†] as Adjudicator in respect of the dispute between you arising under or out of the Contract for …*. Having accepted the appointment and agreed to conduct the adjudication I require your agreement to the following:

[†]*Delete non-applicable words*
*Insert appropriate title etc

1.   The matters to be determined are as set out in the Notice of adjudication.

2.   The adjudication shall be conducted, in accordance with the Institution of Chemical Engineers' (IChemE's) Adjudication Rules ('the Rules') published in 2004.

3.   The rights and obligations of the Adjudicator and the Parties shall be as set out in the Rules.

4.   I shall give reasons for my decision.

5.   The Parties shall be jointly and severally liable to pay my fees and expenses, which shall be paid in accordance with the Rules and the Annex hereto.

6.   I, my company or firm, and IChemE shall not be liable for anything done or omitted in the discharge or purported discharge of my functions as Adjudicator unless the act or omission is in bad faith, and any employee or agent thereof shall similarly be protected from liability. The Parties shall save harmless and indemnify me, my company or firm, and IChemE and any employee or agent thereof against all claims by third parties in respect of the adjudication and in respect of this the Parties shall be jointly and severally liable.

Please sign the enclosed copy of this letter in the space provided, and return it to me within seven days.

Yours faithfully,

Adjudicator's name, etc

Annex: Adjudicator's fees and expenses

The following shall govern the payment of my fees and expenses.

a. For time spent in connection with the adjudication including but not limited to preparation, examination of documents, attendance at any hearings in the adjudication and such other duties as may from time to time be necessary, a fee of ...... pounds per hour exclusive of VAT.

b. This rate is inclusive of administrative overhead costs, including secretarial and local communication costs incurred by the Adjudicator.

c. The Adjudicator's travelling time in connection with the adjudication, other than travel between his residence and his normal place of work, shall be regarded as working time with a limit of eight hours chargeable travel time per twenty-four hour period for any one journey.

d. All other expenses of the Adjudicator incurred in the course of his duties will be reimbursable at cost.

e. Payment of fees and expenses will be due not later than fifteen days after the date of the relevant invoice, and I reserve the right to charge interest daily at a rate of ...% per day on unpaid amounts from the due date for payment of the relevant invoice.

Party's agreement to the above terms

We confirm our agreement to the terms set out above.

Signature: . . . .

Name: . . . .

Position: . . . .

For and on behalf of: . . . .

Date: . . . .

The following shall be this Party's representative in connection with this matter: . . . .

Name: . . . .

Contact details: . . . .

Annex C—Form of application to IChemE for nomination of an Adjudicator

Referring Party's reference number   *Referring Party's address*

                                                        Date

The President
Institution of Chemical Engineers
Davis Building
165–189 Railway Terrace
Rugby
Warwickshire CV21 3HQ, UK

Sirs,

Application for the appointment of an Adjudicator

The application must include the following:

1.   *Contract reference.*

2.   *Form and edition of contract—for example, 4th edition Red Book.*

3.   *Date of the Contract.*

4.   *Copy of the clause relating to adjudication and any alterations to the clause(s) relating to the dispute(s).*

5.   *A copy of the Notice of adjudication.*

6.   *Indication of the type of professional considered to be necessary.*

7.   *An undertaking to be responsible for the payment of the professional fees and expenses of the Adjudicator, including any fees and expenses due if a negotiated settlement is reached before the Adjudicator gives a decision.*

8.   *Payment of IChemE's fixed charge.*

Yours faithfully,

For and on behalf of the referring Party

Annex D–IChemE's list of Adjudicators and charges

A list of suitably qualified persons from which an Adjudicator may be selected by the Parties is available free of charge from IChemE's Rugby headquarters.

A charge will be payable to IChemE with any request for the nomination of an Adjudicator. This payment will cover all work done by IChemE up to and including the Adjudicator's appointment.

IChemE's charge may be ascertained by contacting the Rugby headquarters (telephone +44 (1788) 578214).